★ MILITARY VEHICLES ★

SUBMARINES

MARTY GITLIN

PUBLISHERS

2001 SW 31st Avenue
Hallandale, FL 33009
www.mitchelllane.com

First Edition, 2021.

Author: Marty Gitlin
Designer: Ed Morgan
Editor: Morgan Brody

Little Mitchie is an imprint of Mitchell Lane Publishers.

Title: Military Vehicles: Submarines / by Marty Gitlin
Description: Hallandale, FL :
Mitchell Lane Publishers, [2021]

Series: Military Vehicles
Library bound ISBN: 978-1-58415-785-4
eBook ISBN: 978-1-58415-787-8

Photo credits: Freepik.com, Freevector.com, Cover: U.S. Navy, p. 5 public domain, p. 6 U.S. Navy, p. 7 U.S. Navy, p. 9 public domain, p. 10 public domain, p. 13 U.S. Navy, p. 15 freepik.com, p. 17 Rémi Kaupp, p. 18 U.S. Navy, p. 19 U.S. Navy

CONTENTS

Words in **bold** can be found in the Glossary.

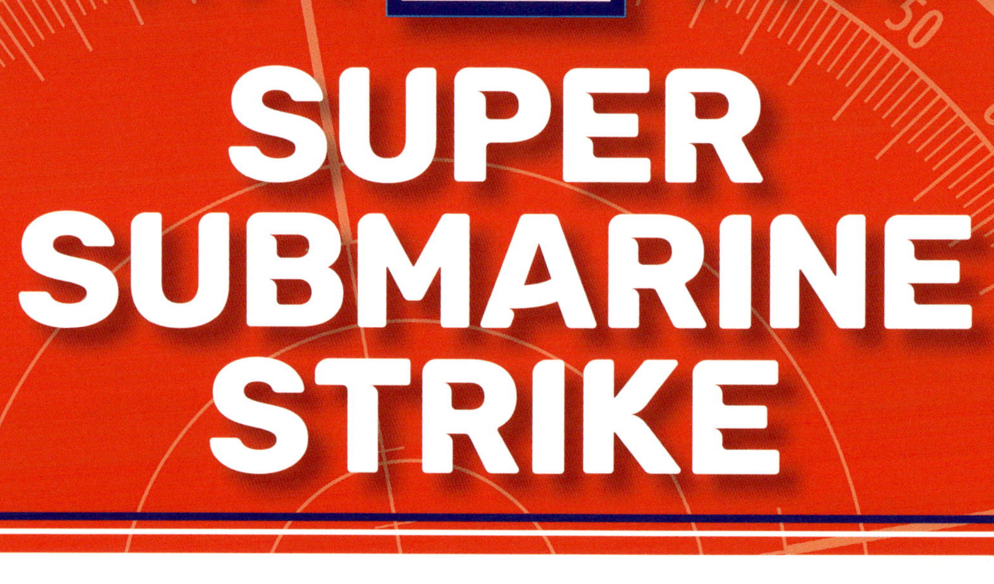

1 SUPER SUBMARINE STRIKE

It was November 28, 1944. World War II was still raging. The fate of mankind hung in the balance.

The U.S. submarine *Archerfish* was on **patrol**. Its crew was looking for Japanese enemy ships. And they spotted one in Tokyo Bay.

That ship was the *Shinano*. It was an aircraft carrier. And it weighed 144 million pounds!

It turned toward the *Archerfish*. Sub **commander** Joseph Enright gave an order. He wanted the ship sunk. The *Archerfish* sent six **torpedoes**. They rushed toward the *Shinano*.

The *Shinano* aircraft carrier was very large.

The big boat had no chance. All six torpedoes slammed into it. The craft exploded. The *Shinano* sunk. It was the largest warship ever destroyed by a submarine.

The sinking helped America control the naval war. The Japanese were soon defeated. The *Archerfish* had played a role in the U.S. victory.

The *Archerfish* is shown around 1945.

The *Archerfish*

FAST FACT

GETTING WHAT THEY DESERVED

The *Archerfish* and its crew were rewarded after the war. They earned the Presidential Unit Citation for sinking the *Shinano*.

2

GOING BACK IN TIME

The idea of underwater craft goes back 2,500 years. One tale claimed a Greek king dropped into the sea in a glass barrel to study fish.

It would take centuries to turn dream into reality. The **Dutch** invented the first working submarine around 1620. It was an enclosed rowboat.

The crew powered it with oars. A sloping front **deck** lowered the sub when they rowed forward.

Early submarines often failed. Among them was the *Turtle*. That U.S. craft tried in vain to sink a British ship in 1776.

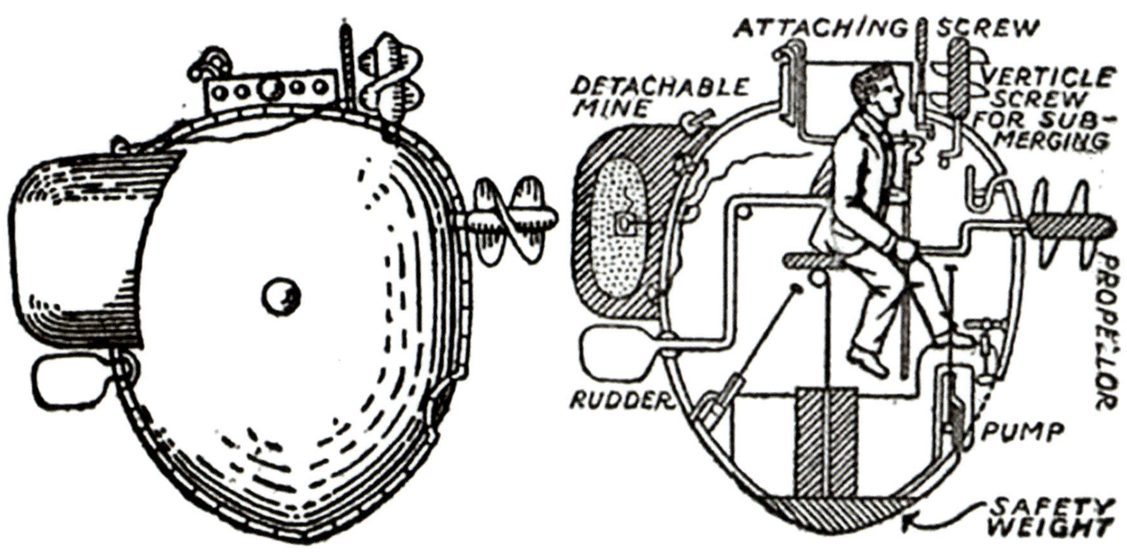

A sketch of the *Turtle*.

The Civil War moved sub warfare forward. The South built one that attacked a Northern ship in 1864. The sub had an explosive device attached to its **nose**. It blew up on contact. Neither craft survived.

That battle launched a new era. Naval warfare would never be the same.

A painting of the Civil War era submarine, the *H.L. Hunley*.

FAST FACT

THAT DIDN'T WORK!

The Dutch built a 72-foot long sub around 1652. It was supposed to ram into enemy ships. But it failed. It could not ram into enemy ships because it was unable to move!

3

SUBS AND SINKING SHIPS

What cannot be detected helps in war. Submarines travel underwater. They often go unseen. They take enemy ships by surprise. They then send torpedoes to sink them.

The *USS Sea Owl* is shown during World War II.

Subs helped the U.S. and British win World War II. American subs sunk Japanese warships in the Pacific Ocean. The British sunk German U-boats in the Atlantic Ocean.

The United States has since rarely used subs in wartime. The military wants them to act as a **deterrent**.

They have done just that since the 1950s. That is when the U.S. and Russia built more dangerous subs. It was hoped the threat they posed would prevent war.

That remains the plan. The U.S. is happy to use subs for other tasks. American subs learn more about the undersea world. Such research can help people in the future.

USS Rhode Island returns to Kings Bay, Georgia in 2013.

FAST FACT

FIGHTING BACK

Sensors are one way to fight back against sub warfare. The U.S. Navy places them on the sea floor. They detect enemy subs. That allows U.S. ships or planes to find and destroy them.

HOW THEY WORK

Submarines are a marvel of science. They can move underwater without rising or sinking.

Subs use the principle of **buoyancy**. They **displace** water equal to their own weight. That allows them to float. They fill their **ballast tanks** with water or air to control movement.

The tanks fill with air when the sub is above the water surface. The weight of the craft is then less than the water it displaces. That allows it to stay up.

The sub releases a vent in the tank to dive. This allows water to rush in and add weight to the craft. That makes it sink. The crew closes the tank when the sub reaches its desired depth.

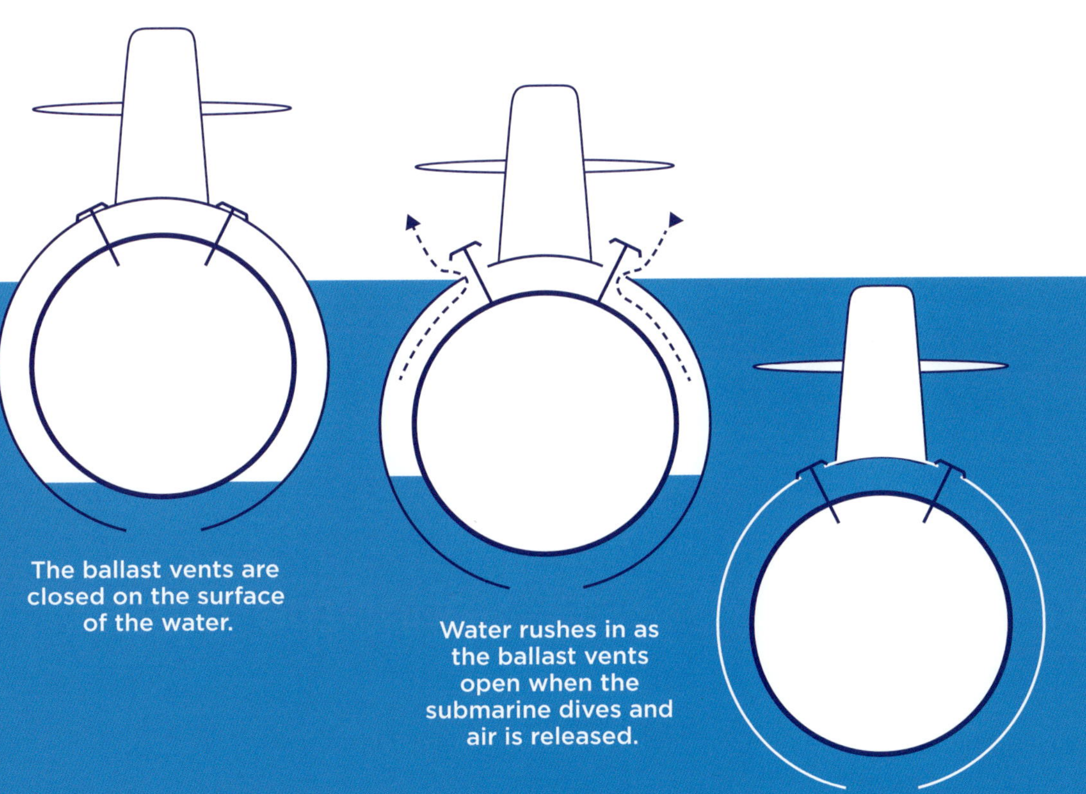

The ballast vents are closed on the surface of the water.

Water rushes in as the ballast vents open when the submarine dives and air is released.

The ballast vents are closed underwater so the hull stays full of water, keeping it below the surface by adding weight.

Military tanks are fitted with many devices. One is **sonar** that detects enemy ships. Another is **radar** that warn against air attacks.

Modern subs are powered by nuclear energy. They also use more than torpedoes as weapons. They can shoot **ballistic missiles** at targets.

A Trident II D5 missile launches from the *USS Nebraska*, a ballistic missile submarine, off the coast of California in 2018.

FAST FACT

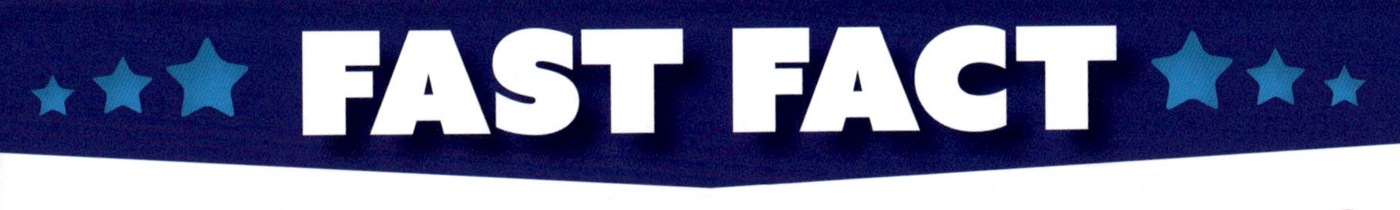

ABOUT THE *NAUTILUS*

The U.S. sub *Nautilus* was built in 1954. It was the first nuclear powered submarine. The *Nautilus* never needed to come up. It was built to stay **submerged**. It became the first sub to reach the North Pole under ice in 1958.

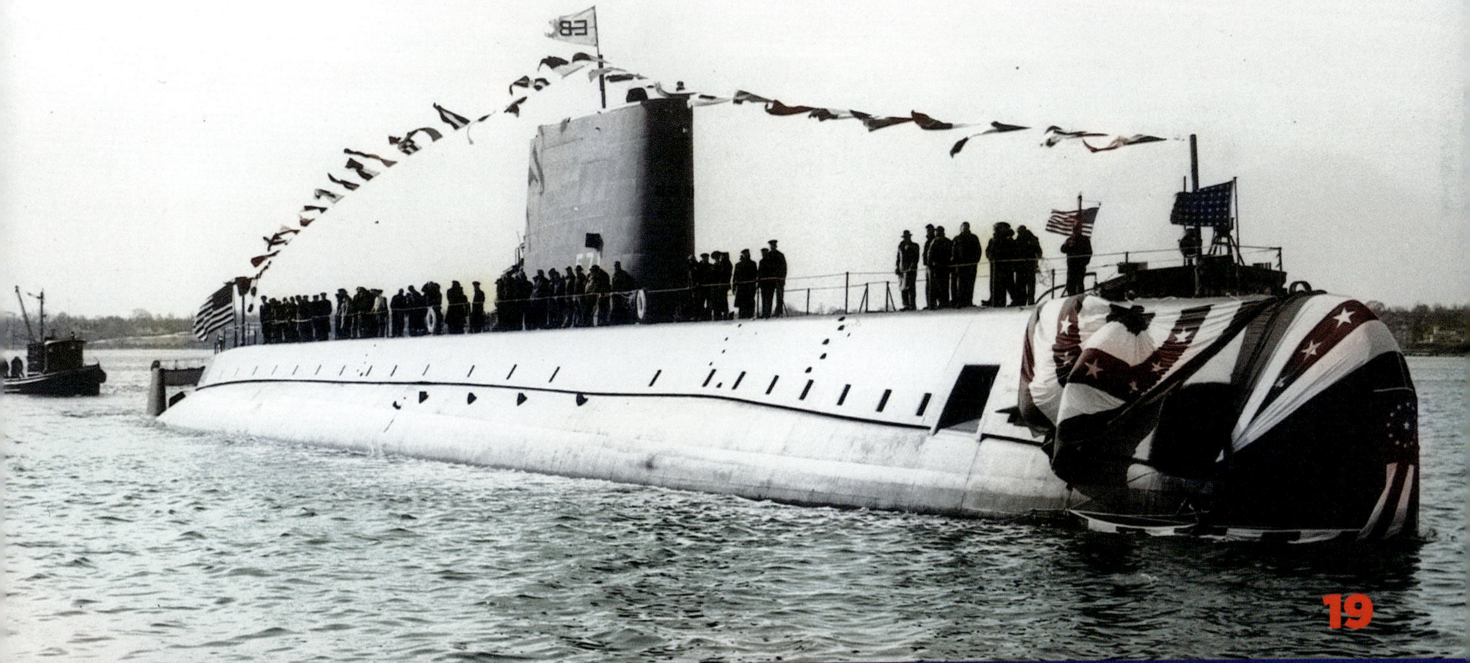

5 STILL GOING STRONG

The use of U.S. submarines peaked in World War II. They have since served mostly as a deterrent to war. And they have been used to explore and research.

That is likely to continue. But the military plans to keep improving its subs. Some of those submarines have gotten old. They must be replaced.

The United States is building new attack subs. They are designed to attack and sink enemy craft. Then it plans to create new missile subs. They destroy targets with long-range weapons.

Submarines can also be used as learning tools. There is much to know about ocean life. Such research can help people and create a more peaceful world.

FAST FACT

THE SUB SCORECARD

The U.S. has 37 attack subs. Another 14 can fire ballistic missiles. Four are fitted with guided missiles.

What You Should Know

- Staying unseen makes submarines most effective in war.
- Sub warfare reached its peak during World War II.
- The U.S. Navy used submarines to defeat Japan in World War II.
- Filling and displacing tanks with air and water allow subs to rise and sink.
- Modern submarines can fire ballistic or guided missiles.
- Subs use sonar and radar to detect enemy ships and planes.
- The greatest World War II sub battles involved the Germans and British.
- The U.S. has rarely used subs in war since the 1950s.
- Subs only used torpedoes as weapons until missiles were invented.

Glossary

ballast tanks
Object that can be filled with or lose water to make subs rise and drop

ballistic missiles
Rocket-powered explosive weapon launched in a high arc

buoyancy
The ability to remain afloat

commander
Person in charge of a military outfit

deck
Floor of a ship

deterrent
Able to stop a bad event from happening

displace
To remove from a usual place

Dutch
People from the Netherlands

nose
Front point of a submarine

patrol
The action of going into an area to observe or guard

radar
Device that sends out radio waves for detecting and locating an object

sensor
Device that detects movement or light and responds by putting out a signal

sonar
Device that detects submerged objects by sound waves

submerge
To go underwater

torpedo
Self-propelled submarine weapon

Find Out More

Capetanopoulos, Demitri. *Ned the Nuclear Submarine*. Zanesville, OH: Proving Press, 2018.

Doeden, Matt. *Submarines (Pull Ahead Books - Mighty Movers)*. Minneapolis, MN: First Avenue Editions, 2005.

Desmond Fox, Tanha. *My Daddy is a Sailor*. Canon City, CO: Lionheart Group Publishing, 2018.

Websites

Ducksters: This site is all about the Navy and other branches of the U.S. military.
https://www.ducksters.com/history/us_government/united_states_armed_forces.php

Science for Kids: This website provides interesting facts about submarines
https://www.scienceforkidsclub.com/submarines.html

Submarine Safaris Kids Club: Learn more about how subs work on this site
https://www.submarinesafaris.com/kids_learn.php

Index

About the Author

Martin Gitlin is an educational book author from Cleveland, Ohio. He has written about 150 books since 2006. Most have been for young students. He has authored many books about military history. Among his strong interests is submarines. Martin spent 11 years as a newspaper journalist. He won more than 45 awards from 1991 to 2002. He earned first place for general excellence from Associated Press in 1996. That organization also voted him as one of the top four feature writers in Ohio in 2002.